Expansion

by
Danny Strack

8

www.dannystrack.com

A Segment of the Universe
(with Cosmic Background Radiation)

The Milky Way
(You Are Here)

Dedicated to People Who Still Read Books
(Like You)

Astronomical thanks to Lacey Roop, Tova Charles & Faylita Hicks
for allowing me to reprint our group poem, "Wishing Well",
in this book.

Intergalactic thanks to Doug Shields for fact-checks and edits on,
"Philosopher Frog Expands My Mind",
and additional feedback.

Infinite love and thanks to
Winnie Hsia, Ruff Draft, Lauren Pool & Robby Cale
for edits, feedback and other contributions throughout the book.

The Milky Way

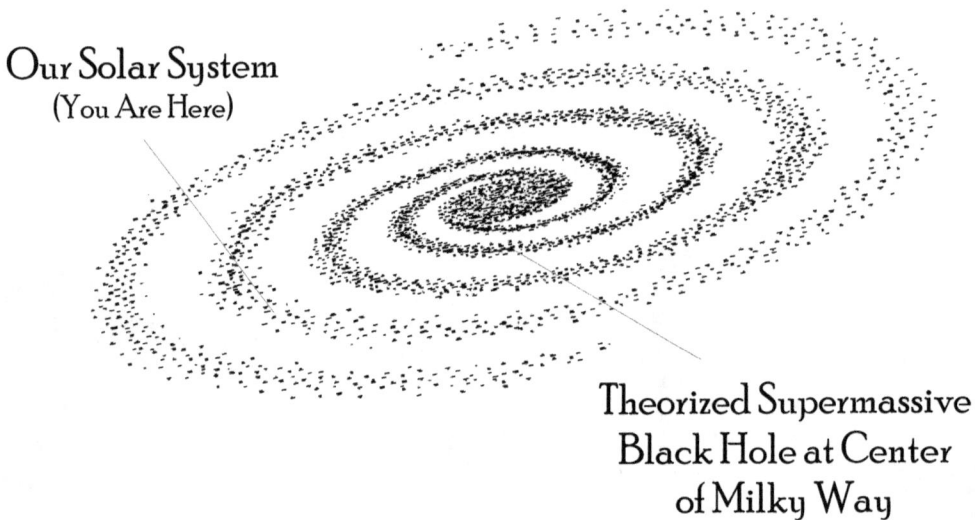

Our Solar System
(You Are Here)

Theorized Supermassive
Black Hole at Center
of Milky Way

—Figure 1—
Three views of our position in the universe.
(Pages 2-5)

Our Solar System
c. 2011, w/ Dwarf Planet Orbits

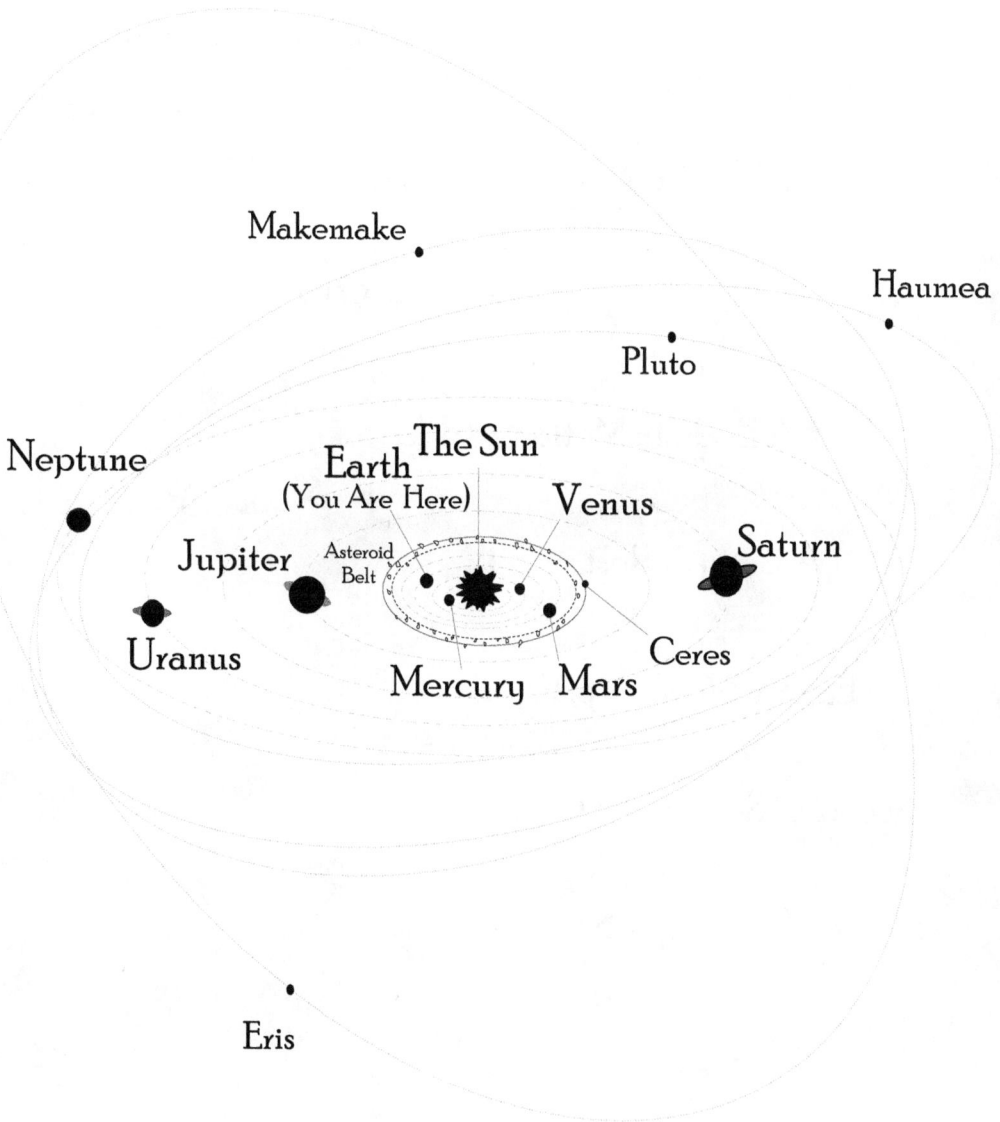

Makemake

Haumea

Pluto

Neptune

Earth
(You Are Here)

The Sun

Venus

Jupiter

Asteroid
Belt

Saturn

Uranus

Ceres

Mercury

Mars

Eris

Map of Contents

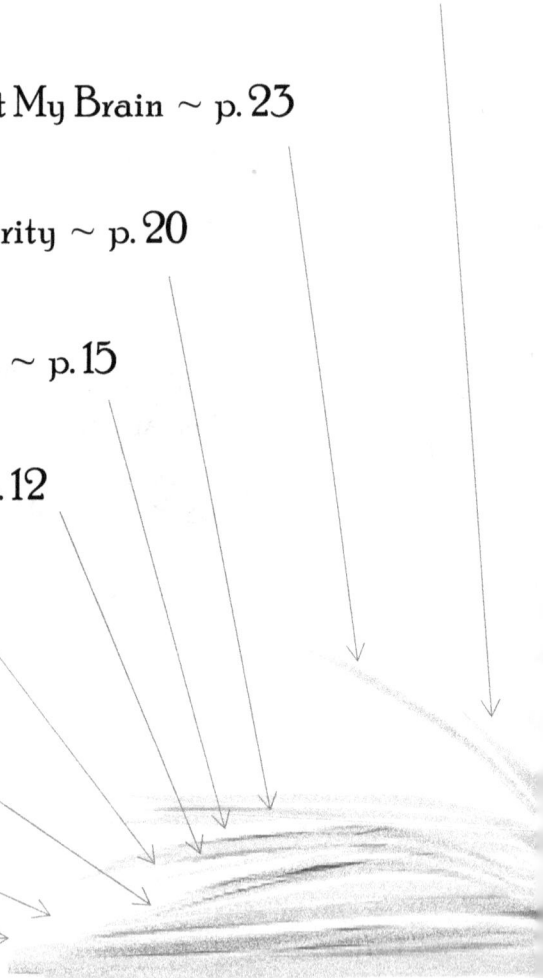

Here on Earth,
we name our stars like sons and daughters,
but the Sun is our father like the Earth is our mother,
the Romans named him Sol.

How appropriate that all our energy came from him.

What souls would we have without the Sun to spark them?

School of Stars

I

Life arcs off the surface of the Sun,
like sparks from a welding gun
shooting molten motion in every direction.
We, the living plant and animal civilizations,
are sprawled out across a bulging round rock,
like lizards basking it in.
Life is where the action is.
If it weren't for us,
the planets spin cool and silent as copper coins on polished granite.
There is no movement on their faces but mechanical wind & moon tides.

II

On our round rock,
every living self is made of living cells.
And every living cell contains a spark of energy.
So we are bursting with sparks.
We would be full of fireworks if you could see in microscopic infrared.
We measure this energy in volts or calories or watt-ever,
but we burn them when we move.
Our fires must be replenished by consuming more sparks.
Everything we eat still flickers with the dying embers of life,
that came from life,
that came from life,
that came from the Sun.

∞

What if our flesh was invisible?
What if we could see at the speed of light?
Could we see what drives us then?
With fires burning like campsites across our continents?
Would we look like lamps?
How bright would we glow?
Could you see us from the moon?
Would there be galaxies swimming in Earth's ocean?
Would we each be a school of stars?

Earth's Blood

Energy incarnate.

Dark as the depths and coal slick.

Compressed from the unquenchable spark in each of us,
from mighty oak,
to the tick-mite on its leaves;
that spark, that flame, that
flickers and dies without consuming and burning
the only fuel everything runs on:
that unquenchable spark in each of us,
from beef to lettuce leaf,
Tyrannosaurus steak to fern frond,
pre-fossil fungus and the blood of invertebrates,
every bit left over after the living had their share
is compressed into energy incarnate,
dark as the depths and coal slick.

Our cars, our trucks, our electric grids,
all run on the pressed flesh of our ancestors.

We burn the blood of our totem animals,
pumped from underground veins with huge syringes.
This is not just the life of our machines,
but the lives of billions of prehistoric plants, animals and algae.

Now dinosaurs are swimming in the ocean again,
and roaming the Louisiana marsh.

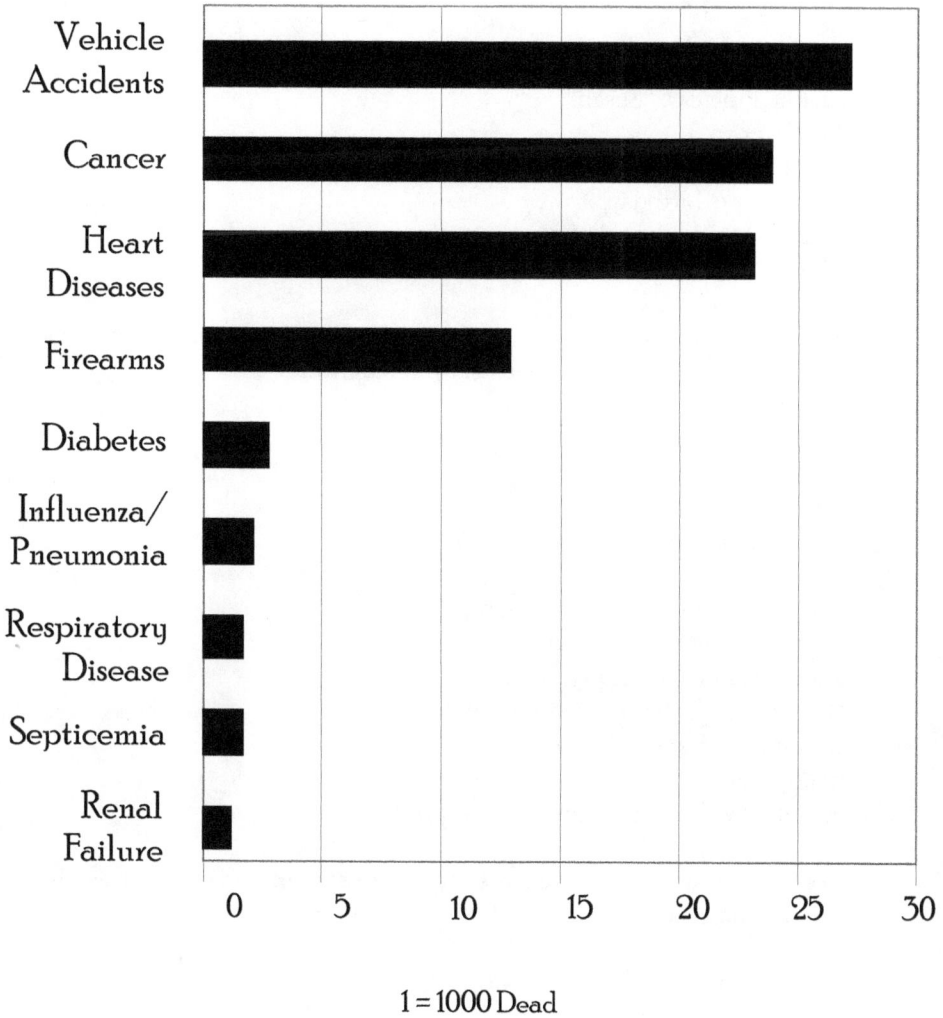

—**Figure 2**—
Most Likely Causes of Death
Americans Aged 0—44

Source:
Mortality Data from the Center for Disease Control and Prevention, National Vital Statistics System
http://www.cdc.gov/nchs/deaths.htm

A Body at Rest

I named my car, "Susan".
My first car,
my father's old Honda,
a gift on my birthday,
I learned to drive in that car,
I loved that car.

For nearly 15 years,
our bodies rolled together over back hill highways
then pinballed up downtown alleys,
my hand the steering wheel, my feet, tires.
And she was so reliable,
started every morning,
I couldn't imagine Susan ever dying on me.

But I didn't always treat her right,
never kept up the maintenance,
always ran her on empty.
Near the end, she smoked too much,
forgot how fast she was going,
made creaks and ticks I never got checked out.
Eventually she got so sick I had to put her down.

Did you know that car accidents are the most common cause of death
for Americans under the age of 40?

Maybe that's why I loved Susan –
I trusted her with my life,
she never let me die.
And death
is terrifying.
I think about it all the time.

I think it was learning of death that first made me conscious I was alive.

Just a little boy, already an insomniac, couldn't keep my eyes closed,
afraid of half-light shadows.

And death, my father explained was not to be feared,
but held as a reminder that every day must be cherished,
lived to the fullest,
because one day I will run out of days.

This deeper understanding of death did not help me sleep.

But it drives me to be fearless freewill freewheelin',
up till morning most nights chasing cloudy daydreams
everywherever they lead.

In my cloudy daydreams,
cars look like turtles,
humans are such fragile bodies in these steel glass shells.
You could crack one just by pressing your foot down too hard.

In my clearsky daydreams,
humans look like turtles,
we have such fragile souls in these human body shells.
You could crack one just by pressing your foot down too hard.

My body is named, "Danny".
a gift from my parents on my birthday,
I learned to live in this body,
I love this body.

He is so reliable,
starts every morning,
even though I don't always treat him right.
He smokes too much,
forgets how fast he's going,
makes creaky ticks I need to get checked out,
I can't imagine Danny ever dying on me,
but worrying about it only keeps me up at night.

When I couldn't stop thinking about death and still couldn't sleep,
my father taught me a new technique to distract me from my fears.

You focus on each tiny part of your body one at a time,
and say goodnight,
then you feel it gently drift away…

Goodnight toes,
goodnight heels,
goodnight souls…
and you feel your tires gently drift away.

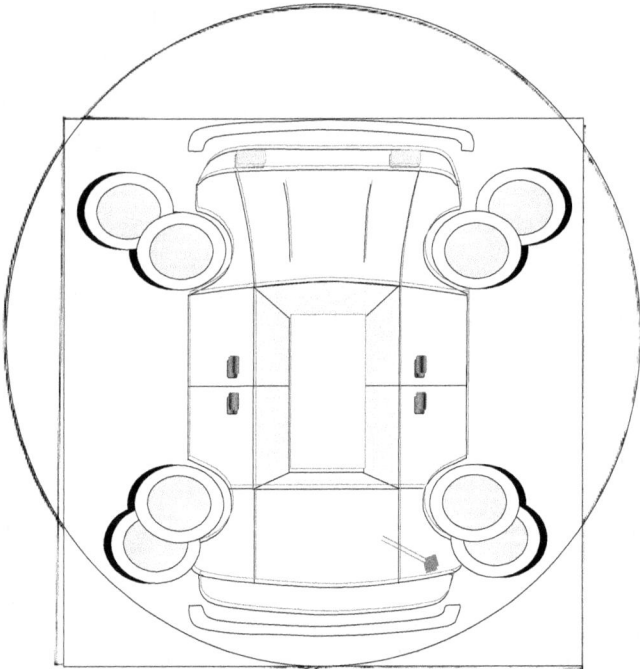

—Figure 3—
Vitruvian Car

A Body In Motion

Parents, Teachers and Girlfriends agree:

I need to stop twitching my leg up and down.
It's distracting,
it is very distracting.
But that's just part of being me,
I'm full of energy,
twitchy, jittery,
always need to be moving up and down,
flexing, flinching,
spinning pens,
waving tongue,
chewing gum,
and sometimes I let my arms and legs totally spazz out.

And that's fine,
It doesn't bother me,
Even when I shake so much I'm shimmery,
I think constantly moving keeps me in pretty good shape.

I don't believe in restless leg syndrome –
never heard of it till last year.
I think a lot of modern syndromes are just excuses to help sell pills.
They slapped a label on twitchy people so they could feed them downers...

Though maybe there's also something to be said for predisposition.

I remember
when my grandfather was still alive,
and we went into restaurants,
he would always order his glass of water half-full...
Not because he was trying to illustrate his optimistic outlook on life
at every meal...
but because he couldn't stop his hands from shaking,
and he didn't want to spill.

Plagued with tremors as long as I knew him,
slight at first...
Grandpa could still bullseye in the backyard beanbag toss,
and whittle new walking sticks for our hikes in the woods,
and bake the most amazing bread –

He fought in World War II but didn't really talk about it.
He'd rather tell stories from after he left the military,
discovering his long lost family in his homeland of Italy.

When he got back to the states,
he married my grandmother and adopted her Quaker faith.
He was a pacifist, teacher, and social worker for the rest of his life.
He taught sex education before it was buzz word,
mentored the youth and mentally challenged.

By the time I met him,
he was a perfect grandfather with a handsome gray beard,
tall like me,
and he always made me feel proud of myself,
and gave me awkward advice for picking up girls
when I was too young to understand.

But maybe it's things I don't remember that are more important than
the anecdotes.

I never heard him raise his voice in anger,
never saw him angry,
or frustrated,
even when he shook so much he was shimmery,
he never let a day pass without laughter,
and he never took my grandmother for granted.
He loved her with a fiery Italian passion,
even as the tiny flames of Parkinson's burned down his spine,
he loved her deeply until the day he died...

So if this is the tree I was carved from,
then give me its grain.

If this is the man I could one day be,
I would gladly share his disease.

For if the hand fate dealt me is one that shakes,
I will hold that hand at the end of my wrist,
and with joy I will drink from my cup half full,
and do my best not to spill.

---

1. A body at rest remains at rest unless it is acted upon by an external force;

Likewise, a body in motion remains in motion and moves along a consistent and continuous path unless it is acted upon by an exterior force.

~

2. Force equals mass times acceleration.

I.e., $F = ma$

~

3. For every action there is an equal and opposite reaction.

—Figure 4—
Newton's Principle Laws of Motion

(1)
Cosmic Radiation
Particle / Antiparticle
Pairs Are Created
in the Gravity Wells
of Black Holes.

Gravity
Well

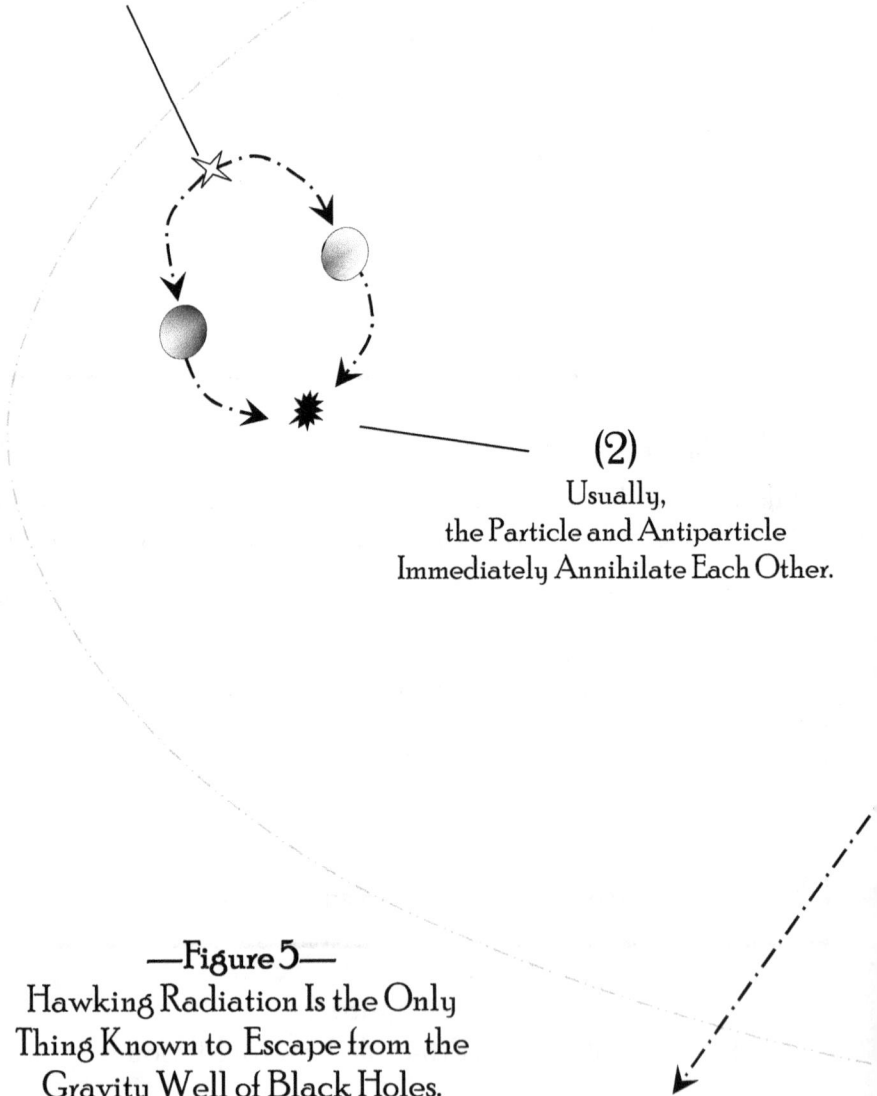

(2)
Usually,
the Particle and Antiparticle
Immediately Annihilate Each Other.

—Figure 5—
Hawking Radiation Is the Only
Thing Known to Escape from the
Gravity Well of Black Holes.

(3)

However, When a Particle / Antiparticle Pair Splits on the Edge of the Black Hole's Event Horizon...

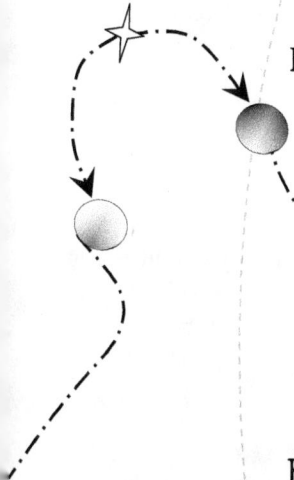

Event Horizon

(4)

Half Enters the Black Hole.

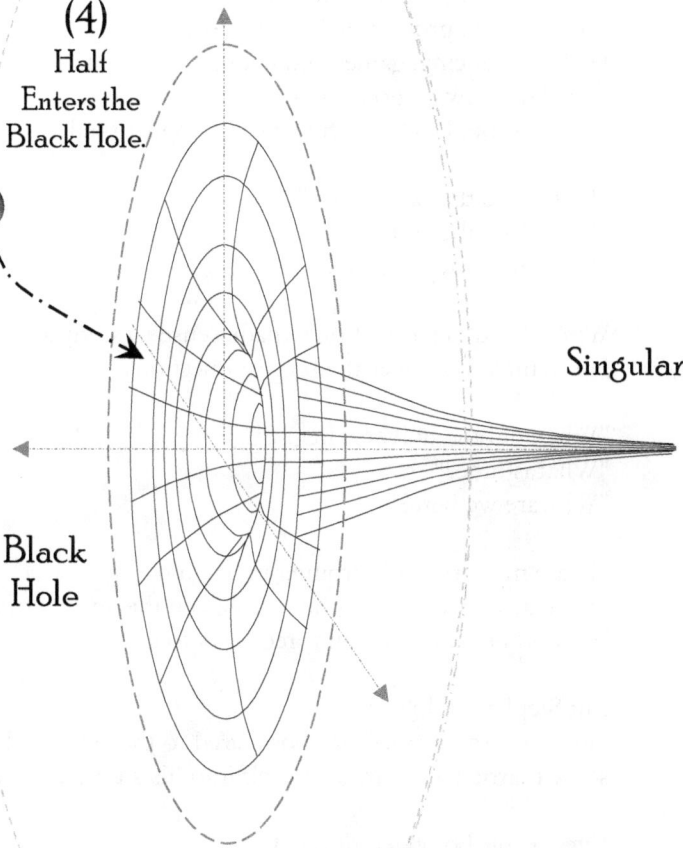

Singularity

Black Hole

(5)

While the Other Half Escapes as Hawking Radiation.

Sources:

T. E. Bunch & J. H. Wittke. "Meteorite Glossary, Hawking Radiation.", Northern Arizona University Department of Geology, Northern Arizona Meteorite Laboratory , Accessed Dec. 21, 2010. http:// www4.nau.edu/meteorite/Meteorite/Book-GlossaryH.html

David Shiga. "Hawking radiation glimpsed in artificial black hole." New Scientist, Sept. 28, 2010: Physics and Math, Accessed Dec. 21, 2010, www.newscientist.com/article/dn19508-hawking-radiation-glimpsed-in-artificial-black-hole.html

Singularity (Remains in Motion)

Born in Oxford during the blitzkrieg on London,
little Stephen loved to move.

He had 11 ways into his house –
not all on the ground floor.
He liked dancing, games and taking things apart,
they didn't always go back together.
But he wanted to know how everything worked:

"How does the radio play?"
"How does the sun shine?"
"Why are some stars blue?"

With a head in space it only made sense to study astronomy in college.
He wanted to answer the biggest questions:

"When did the universe begin?"
"What is time?"
"Why are we here?"

These mysteries still orbiting in his mind,
when at twenty-one Stephen was given the answer the ultimate question,
and told he had less than three years to live.

But Stephen didn't care.
His problems seemed tiny compared to those posed by the cosmos,
so he married and threw himself into his studies.

Even as his body deteriorated,
his mind took off like a rocket.

He developed theories on singularities in general relativity,
contributed to our understanding of cosmology and quantum entropy.

And he defied his doctor's orders when he refused to die,
still alive a decade later but completely paralyzed.

And even though he couldn't move,
his mind wouldn't pause.
He fact-checked Einstein and found flaws.

He developed his own set of mental tools
that made the other physicists look like they were still using slide-rules.
He figured out what caused the Big Bang,
and changed our understanding of time and space
when he proved how radiation is displaced in black holes.

And then he –
com
plete
ly –
lost the ability to speak or any way to communicate.

Black holes create massive gravity wells
that absorb everything within their wake,
from which nothing can escape.

Stephen spent nearly a year in that state.

Brain taking in everything around him in space.
Eyes peering out from the event horizon of a black hole –

Until the day they invented a speech synthesizer and sent it to him,
so they wired new hardware to his wheelchair
and gave Stephen his new voice.

He speaks by writing with the last functioning muscle in his face.
At five words per minute,
he's published five books.
Delivered hundreds of lectures all over the world,
continuing to travel and teach,
and study and research,
never giving up his search for answers,
still alive now forty-eight years after his fatal diagnosis.

Doctors believe he has a special form of his disease.

I say Stephen is a special kind of man.

He doesn't curse his disability,
he considers himself lucky,
defying death for fifty years,
to face down the biggest questions anyone can think up.

Even as he sits,
stuck behind the eye of an event horizon,
Stephen imagines radiation particles splitting in the gravity well,
and uses his last functioning muscle,
to smile.

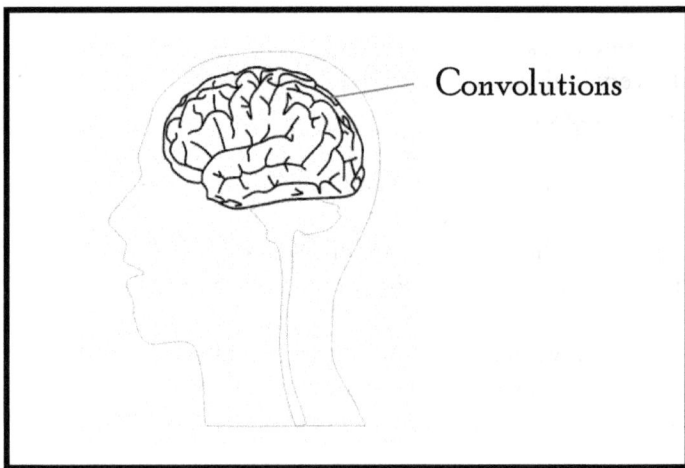

Convolutions

—Figure 6—
con·vo·lut·ed ~ adj.
1. Twisted; Having many overlapping coils; i.e., a hair braid.
2. Complicated, intricate language; roundabout logic.
3. Pertaining to the numerous folds and ridges
on the surface of the brain.

Thinking About My Brain

I know this is convoluted,
but I've been thinking about my brain.

It's convoluted.
Like a pile of thick chain.

I pick up a single link,
and find it attached to a whole web of paths
I can follow back into the hollow of my head.

I have trouble unwinding.

It's been a while since I saw enough daylight.

I have trouble sleeping.

Maybe I prefer my mornings late at night.

Or maybe I can't stop thinking once I start,
but lately I think I feel more with my head than my heart,
always analyzing these emotions instead of experiencing them,
trying to observe all my actions from outside of my eyes.

I don't want to get caught in a mistake,
and besides,
I like to change shoes with every step I take
in the hope these shifting points of view
will give me new perspectives on the best ways to live.

But the side effect is,
I ended up emotionally existential.
Living beside my self instead of inside.

Because it's much easier not to be affected by a life unfolding across an unfeeling gulf,
like a second hand story on a jaded TV screen.

Like I've dammed the mighty river of experience,
and only let a trickle flow through my stream of consciousness.

And I know all this is really convoluted,
but I've been thinking about my brain.

And I'm tired of only drinking diluted moments from the drain.
Whether I face joy or pain,
I want to feel the full weight of life's rain pound my skin.
I want to sink into alive until I learn to swim,
and when I come up for air,
I'll be laughing, crying, shouting together.
Swirling in a whirlpool of unmuted emotion,
reeling in a torrent of feeling,
as waves of sensation crash around my flying form,
I'll find peace no matter the weather.
Arms open,
embracing,
life's beautiful storm.

Wishing Well
(w/ Lacey Roop, Tova Charles, & Faylita Hicks)

We were kids tossing copper in the water,
trying our luck with discarded change,
placing our best bets on each well we passed...

We thought our wishes would all come true:
prayers for world peace and ponies,
flying or having a dinosaur zoo!
they only cost a penny,
a shooting star,
a broken eyelash,
dandelion fluff.
We believed wishes were free.

But then we grew up.
And it started to feel like the price of well wishes was growing too.
Like God put a toll on better tomorrows,
as if you had to be rich to reach your goals,
and every wishing well must be filled with 50s and 100s,
because you didn't just need money to make money...
You need money to make dreams.

But money fades as quickly as the wish you forgot you made.
Good luck and joy can't be bought, sold, or paid.
Granting wishes is far more expensive than we ever imagined.

You can wish all night for the stars to ignite,
and guide you through the dark to the start of a new day...
Or you can build your own fire for light.

Tomorrow is no promise.
So why waste your todays,
subtracting minutes from hours,
marking years off calendars?

Wishes will only come true if you push them through.

The greatest tragedy of all is not failing,
but living your whole life
never trying to complete whatever it is you believe
would make you happy.

If you think your wish is wilting, then cut it in half,
bury part in the sea, burn the other to ash,
metamorphosize it into possibility.
Never forget,
our wishes have always been more valuable than
the fires we burnt them in,
or the whispers that got swallowed by the wind.
So if your dreams have disappeared
then get the rake and scrape them from your skin!
Build yourself a castle with the leftover ash.
Take up the wind sown seeds and nurture them.
Let them grow into gardens of goodwill and harvest them like hyacinth;
to make your dreams come true,
sometimes you have to work until the bone breaks through.

Then your wish will dance,
optimism will pay you back for all the pitfalls that have befallen you.
If you have doubts then serenade them out.
Impress the God of mediocrity with your two step.
There are angels bulging in your veins
Let your scapulas sprout their wings!

So I constructed you a compass from shipwrecked seashells,
in case you forget that even the discarded, lost or forgotten
still have something
beautiful inside to see.

May you continue to be
the person you've always wanted to become.

Because the copper in a coin,
doesn't compare:

to the wish you don't think matters,
to the wish you're trying to forget,
to the wish you've been trying to drown,
that refuses to sink.

~

Thanks again to Lacey, Tova & Faylita
for collaborating on this poem
and allowing me to reprint it in this book!

Haiku

The moon would be invisible
without the sun's light.
And so would we.

~

Worthless unrepairable satellites
make beautiful shooting stars.

~

Driving to work
annoyed by rainstorm.
I pass a farmer thanking God.

~

Lightning flashes in sky.
UFO remains hidden in storm clouds.

~

Stars are simply fireworks.
Though they take longer
to turn ash and fall.

Earth

At first there was earth.
> *and air,*
> *and water,*

Life
gurgled up from the mud.
> *bacteria,*
> *algae,*
> *fungi,*

>> *we are insects,*
>> *we are fish,*
>> *we are vertebrates,*
>> *we are ferrets,*
>> *we are human,*

Whether you believe in evolution or the bible or meteors,
> we all agree on one thing,
> we came from the mud.

from the earth,
of grass and tree,
of mountain and cave,

> We mine earth for minerals.
> *for bronze, for tin, for iron, for lead,*

>> Make steel.
>> *forge shovels,*
>> *forge axes,*

> Chop down trees.
>> *build houses,*
> Plow land.
>> *build fences,*

> Claim land.

Fight.

forge weapons,
forge armor,

Fight.

> Mine nickel.
copper, silver, gold,
Forge money.

Fight.

Land means money.
Money means work.
Work means time.
Time means land.
claim land,

Fight.

Mine sodium. *silicon, sulfur,*
Find fossil fuels.

Fight.

make paper,
make plastics,
make toys,
make lasers,
launch rockets,
launch satellites,

Fight.

From found sticks and hurled stones we forged our society.
factories, laboratories, complexes,
Whatever we could pull up from the mud and process,
time means work means money,
we processed
money means mud means dirt,
into whatever we could imagine.
hypothesize, research, experiment on,
Everything on Earth
prove theories, build bombs,
rose up from the earth and the ocean,

Fight.

like the trees,
chop down trees,
like mountains,
plant flags,
like a child standing for the first time.
or a calf in a veal factory
taking their first steps
brand new leather shoes,

30

As you grow,
down comforter, new moleskin, fresh pen,
you discover you're stuck to the ground like you're made of magnets.
of oxygen, carbon, hydrogen,
Planes trick you into feeling free,
nitrogen, calcium, phosphorus, potassium,
but really you're still stuck in the same inside out fishbowl as ever,
this is what you're made of
with a solid core and thin layer of breathable air on the outer edge.
you and everyone and everything else
Fight.

**the only way to really escape is to die in space
and to leave your body there**

Otherwise you
and everyone and everything else
will return to Earth one day.
wheeze one last collective breath,
wither away,
to decompose
ground back to dust,
back to mud,
back to earth

The Economy Is Down

Oh, hey everybody, it's me –
The Economy!
What's up?
Not me!
Yeah, Economy has been feeling pretty down lately,
and everyone keeps saying that the only way to fix a down economy is to
stimulate me.

Like I've got blue balls instead of blue chips,
and all Economy needs is a hand job to get everybody off –
welfare.

At least a hand job is a job.
Someone ought to be employed.

Let me tell you something:
After the housing bubble burst a few years ago,
Congress passed that first stimulus package for $152 billion,
and it was painful.

But just like kidney stones, babies,
or passing anything else through an uncomfortable orifice,
it didn't hurt nearly so bad the second time in '09
when we passed an additional $185 billion,
Then $400 billion, and another $200 billion!

Economy is now very stimulated!

Economy starting to have trouble retaining upward growth!
I remain small and flaccid and unable to satisfy demand...
There is a lot of pressure being put on me to perform.

The Economy is actually very sensitive if you get to know me.
Economy likes walking in rain and petting kittens.

And above all else,
Economy demands honesty from anyone that wants to be my partner.

Hey, Economy "gets it."
Economy needs to regain your confidence,
but I think really this is more about my confidence in you!
Economy just wants to trust people again.
We need to rebuild our relationship on bonds of mutual trust,
rather than bonds, mutuals and trusts.

Just give Economy one more chance
and boomtime will be back again, baby!

Get ready to show me your gross domestic product,
'cause I'm about to inject liquidity into the marketplace,
and we will exchange commodities!

But first, two things:
Quit buying shit you can't afford on credit.
Also, Economy loves you but, very tired, Economy needs some rest,
before you fuck me again.

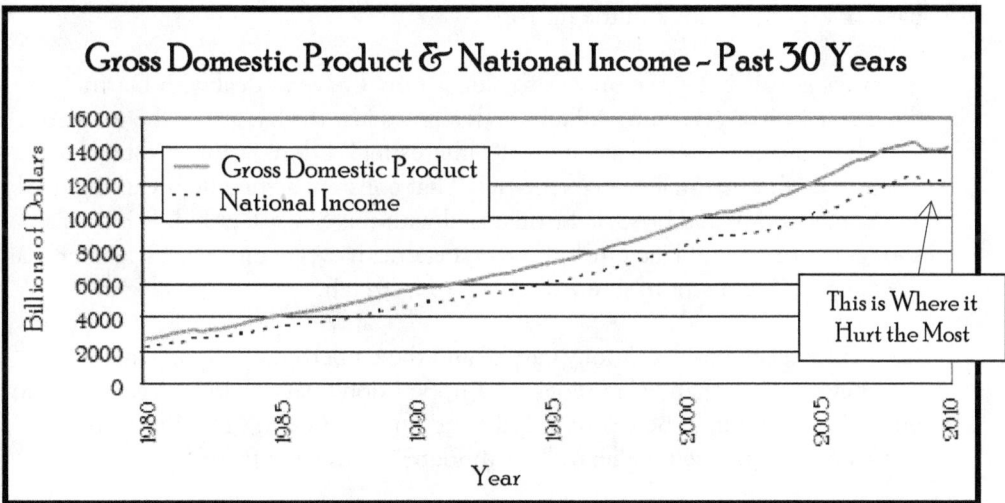

Gross Domestic Product & National Income ~ Past 30 Years

—Figure 7—
The Recent Economic Downturn

Source:
Tables of The Bureau of Economic Analysis National Economic Accounts,
http://www.bea.gov/national/nipaweb/Index.asp

Agent of the Empire

Hello My Dearest Sugarwookie...

It is I, Admiral Piett! "Admiral?" You ask? That's right! Your husband has just been promoted to the newest (and best) Admiral in the Imperial Fleet! And it all happened so quickly... Just this morning we were coming out of lightspeed and my boss's boss, Darth, he's basically Vice-Emperor and a total butthole. Anyways, we come out of lightspeed over some ice planet and Darth calls up my boss on the screen and strangles him to death right in front of me! Then he promotes me to Admiral, puts me in charge of the fleet, and leaves me to call the droids to take care of the corpse. It was awesome!

And here's the best part: The reason Darth was so pissed, was I "forgot" to pass my boss the message about how we were supposed to stay in hiding. So the rebels were able to put up a fight and escape from their base, but it doesn't matter to me, 'cause now I'm an Admiral.

Oh and Spacetriscut, you wouldn't believe my new private bathroom! Everything is a droid! The sink, the shower, even the toilet wipes up for me! Plus there's this R2BJ unit... And I've been getting all kinds of complements on my new hat... Every time I look in the mirror!

Anyways Banthabuns, the only downside is now I have to deal with Darth directly. Already today, he's called me five times just to chat about the weather... Oh yeah, another asteroid storm, really interesting! I don't want to sound insensitive, but he totally creeps me out! That constant asthmatic breathing. He's a religious nutjob and he's totally obsessed with this Skywalker kid. I think it's some kind of pedophiliac crush, I mean, he already wears more black leather than the gimp in Pulp Fiction, so it isn't much of a stretch.

It must be related to that kooky cult he and the emperor are the only members of. They're worse than Scientologists! I hope I don't have to join to get a promotion. Though it might be worth it. I like the ring of "Grand Moff Piett", but I think I could aim even higher... "Darth Piett?" "Emperor Piett?"

We could move the whole family up to the Death Star! Even your mother! The place is huge, really nice and completely impenetrable! Trust me, C3QT, you are going to love the Death Star!

How to Feel Good About Saving the Environment

Next time you're at your mom's house,
and you go in the bathroom to masturbate...
Don't leave the water running to hide the sound.

It wastes water!

Don't worry about your mom,
worry about Mother Earth instead!

Even if you're using a vibrator.
Just tell mom it's an electric toothbrush,
and you are really anal about dental hygiene.

And I hope you're using rechargeable batteries in that vibrator.
Regular batteries release mercury, lead and cadmium into our environment.

Personally, I can not orgasm if I know I'm polluting at the same time!

That's why I always masturbate in the dark,
with the A/C off,
to save electricity.

You don't need light or cool to touch yourself!

It should be dark, sweaty and shameful!

And you really shouldn't clean up afterwards either.
That wastes water, soap and paper.

Mother Earth wants you to sit in it.

Mother Earth

Like a teenage daughter,
who never even considered what she would do with kids of her own,
Earth is too young to be our mother.

She was only experimenting when we were conceived;
wild, reckless and young,
after tame fields and herd animals –
the birth of our civilization was a mistake.

We, wondrous might-be-mistakes,
we child prodigies,
simply intricate,
of laughter and joyful symphonies,
perfect algorithms and profitable companies,
we overachieving burdens on our unready mother,
with endless excuses for our unmade beds,
we take all she offers and ask for more.

We grew like weeds.

She often sees us as angry brooding children,
playing with stolen fire and atomic bombs,
she fears we might burn down the house one night.
She would give up,
but has already sacrificed so much of her body to raise us,
stretched beyond her resources to put food in our bellies,
and clothes on our backs,
she is so easy to take for granted.
We hardly notice her calls –

We hardly notice our mother,
of the cavernous voice and mountain range bosoms,
of seven thousand seven seas and millions of tiny islands
freckled with rainforests of giving trees.

We of the dynamite and chainsaws,
of endless stretches of road,
of cruise ships and cruise missiles.
We came to build something more perfect,
with such beautiful visions for the future,
we take such terrible actions to achieve them.
We fear we might destroy our mother in the process.
We say we want to save her.

But she's been sucking down sunlight a lot longer than we have,
and she'll still be spinning long after we fade away.
We don't need to save the planet –
We need to save ourselves.

We are so small and fragile,
she could stomp us out so easily.

This is no human mother.
Earth is young, wild and unbound by civil law,
and no matter how much she loves us,
if we cut her too deeply,
her reckless whimsy may deem us unworthy and she will strike back.

The temperature will rise like any body fighting a virus.
It only takes a few degrees to melt the ice caps and shrink the land,
or the plates could shift and open a single super-volcano,
to spew forth ash to cover the sky,
or maybe we'll destroy ourselves and save her all that trouble.
Either way,
it wouldn't take much to create an environment unsuitable for us,
but perfect for whatever fresh creatures spring forth from the depths
to evolve into her next new favorite children.

It is the saddest day when a young mother contemplates drowning her kids.
When she can care for them no more.

My fellow children,
it is time for us to grow up.

But What About the Mutants?

"Toxic chemicals!"
They say,
"Nuclear waste!
Global warming!
Landfills, oil spills,
clean cut trees and PCBs!
Human encroachment on wildlife habitats!"

They're screaming that from mountaintops,
"The cute, cuddly animals,
they're endangered,
they're dying,
they're going extinct!

We are responsible,
it's all our fault!"
They try to make you think...
And okay...
It's probably true.

But there's something else they aren't telling you,
there's more to the story than this,
maybe some species of penguin are dying,
and frogs, mice & elephants...
But what about all the new species constantly created?
What about – the mutants?

Mutants swim in toxic chemicals!
Mutants love the heat!
Mutants dine on radiation – it's their favorite thing to eat!
But they also down garbage by the barge,
and suck up oil slicks,
then for dessert they eat latex
and non-recyclable plastics.

In general mutants are always best suited
for habitats that are the most polluted.

Which is perfect,
cause that's the direction things are going anyways.

And who are you and I to say that old animals are better than new?

Why do we think elephants are so great?
Maybe it's just their time to die to make room for the mutants!

And sure, maybe most tomato-fish aren't as cute as squirrels,
but the fact is,
they're just better adapted to this world after all the crap we did to it.

And if the mutants hold any sway,
maybe someday
our whole planet will be that way.

So raise a glass to the pentapus,
the unipede,
and tulip-wheat-dog.

To the spider-eyed-gecko,
and the flying mushroom log.

Out with the old and in with the new
from the feathered gill hippo,
to the day-glo slime goo!

Chasing Elephants

"Elephants!"
Eddy bursts through the door and yells again,
"Elephants! Danny, there are elephants walking down Rosewood!
Come on, we have to see them!"

And I am instantly a naive child again,
transported back in time by my brother's excitement.

I believed everything when I was little.

But then I snap back to adulthood,
I've been fooled before by people crying out about imaginary animals.
It's midnight and we live in downtown Austin.
There are **not** elephants walking down the main street outside our house.

"Elephants?" I ask skeptically, "really?"

"Seriously! I think they're walking circus elephants into town in a parade!
I think I saw an elephant's tail! Come with me! We have to chase them!"

But still I wait,
unwilling to accept a reality in which elephants walk down downtown city streets.

Eddy stares into my disbelief with dismay and says, "Well, I'm going to chase
elephants. You don't have to come." Then he turns and runs back outside.

I look back at the TV I forgot I was watching
and consider how wise I have become with age.

Then I reconsider,
lace up my shoes,
and race out the door,
to run beside my brother and chase elephants.

Whether or not they exist.

Humans
(You Are Here)

Chimpanzees
Gorillas
Monkeys
Primates
Pigs

Dolphin
Whales
Seals
Hippos

Pterodactyls
T. Rex
Stegosaurus

Mice
Rats
Squirrels
Beavers
Porcupines
Rabbits

Birds

Dinosaurs

Crocodiles

Blue Birds
Penguins
Chickens
Hawks

Owls

Snakes
Turtles
Lizards

Lions
Tigers
Foxes
Dogs

Bears
Cats
Wolves

Rhinoceros
Elephants

Horses
Gazelles
Armadillos

Mammals

Doves

Reptiles

Frogs
Toads
Newts

Opossums
Kangaroos
Koalas

Marsupials

Vertebrates

Fish   *Amphibians*

Crustaceans

Oysters
Snails
Squid

Mollusks

Invertebrates

Triggers
Clown Fish
Coy

Fireflies
Butterflies
Dragonflies
Bees
Ants

Insects

Arthropods

Sponges

Bryozoans

Animals

Starfish
Brachiopods
Round
Worms

Spiders
Tick Mites

Arachnids

Coral
Ciliates

Entamoebae

Plants

Fungi

Yeast
Mushrooms

Moss

Flowers
Lettuce

Ferns
Algae

Cherry
Trees

Eucaryota

Protists

Oak Trees

Slime Molds

Trichomonads

Diplomonads

Halophiles

Methanosarcina
Methanobacterium
Methanococci

Thermococci
Pyrodicticum
Nanoarchaeota

Archaea

Green
Filamentous
Bacteria

Aquifex

Thermotoga

Gram Positives

Planctomyces
Bacteroides
Cytophaga
Cyanobacteria

Proteobacteria

Spirochetes

Bacteria

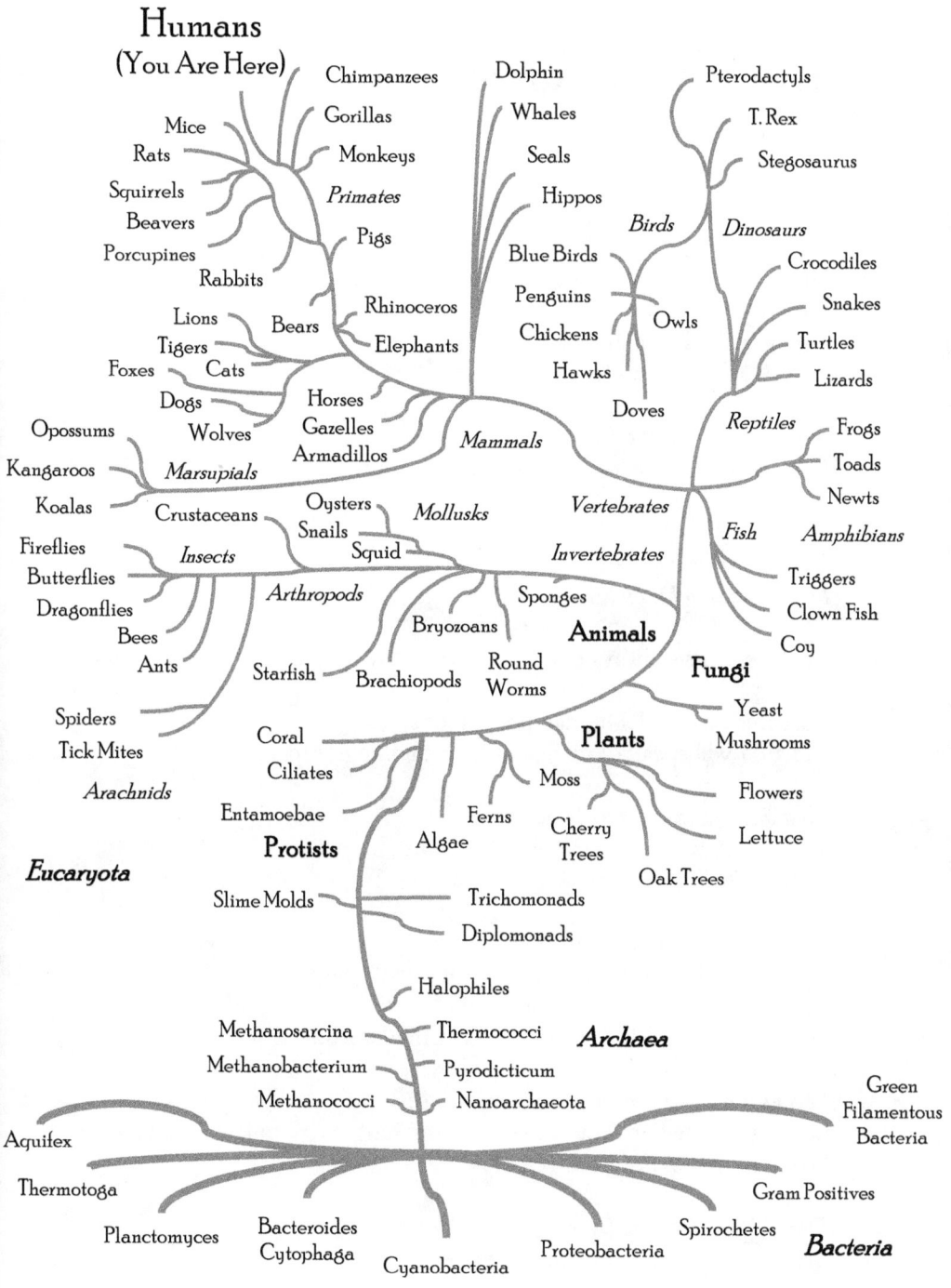

—Figure 8—
A Tree of Life

41

The Mouse Race

Death1 = Death of the Living Spirit

Hi.
I'm Dr. John Calhoun,
I study mice at NIMH, [1]
– the National Institute of Mental Health.

And based on my findings,
I am quite concerned about the mental health of our nation,
and indeed the entire human race.

Death2 = Death of the Living Body
Death1 leads to Death2

You see, I have always felt a great sympathy for small helpless creatures.

I use mice to model human populations,
and so, I will speak of mice, but I am thinking of men.

Most likely causes of mass Death1:
famine, epidemic disease, mass migration, war,
A.k.a., The "Four Horsemen of the Apocalypse".

So I created a mouse universe!
I provided them with all the food they could eat,
I protected them from predators
within my 10 ft. tall, barbed and electrified arms.

I sought to create mouse heaven, you see,
but instead they bred out of control and rang the population bell curve;
boxed in by the Four Horsemen.

Death Squared = The Four Horsemen
Most likely cause of Death Squared = species overpopulation

You see, all that separates mice from men is 5 percent of our genes.
Opposable thumbs and convoluted brains –

About the same amount of hair on much bigger bodies.

And no one kills humans anymore except ourselves
and we have plenty of food,
so our population is growing,
even if we can't see the effects directly yet,
we understand all this is happening.
It would be easy to imagine man sinking into the same hole as my mice.

Death Squared "leads to dissolution of social organization...
=loss of capacity to engage in behavior essential to species survival" [II]
= mass Death1
= mass Death2
I.e., Death Squared = species extinction.

But this need not be the case with men!
We are not mice! You see?
We need not fall at the Horseman's sword!

That is the real difference between mice and men –
not just thumbs and brains,
but our ability to learn and change!

Survival1 = Survival of the Spirit
I.e., Experimentation = Understanding = Compassionate Action in Reaction to Problems
I.e., Innovation.

In my experiments,
my mice sometimes make discoveries that support the entire group!
Ways to make difficult tasks simpler and improve quality of life!
Some mice basically discover the wheel!

And these ideas never come from mainstream mouse society, you see,
but rather from those withdrawn in the corners of my universe,
"the disorganized," (awkward, queer) "subordinates" [III],

Therefore,
we must encourage "deviant, creative, and thus adaptive behavior"
as the means to ensure man's survival!

Survival2 = Survival of the Body
I.e., Innovation = Creation of New Intellectual Spaces = Intelligent Resource Utilization
= Mass Survival1
= Mass Survival2

Based on my research,
I have designed mouse colonies that stimulate these "creative deviants,"
and heal the spiritual death of the mice!
My communities could last indefinitely,
and we could do the same.
This is the real secret of NIMH!

An educated population = natural reproduction stability over time
I.e., Education = empowerment through information,
I.e., wisdom is the key to our salvation.
I.e., nutrition, conservation, contraception, equal rights.
A.k.a., the four horsemen of the revival.
A.k.a...,

...Survival Squared!
You see!?!

Sources
I.   Henry Fountain. "J.B. Calhoun, 78, Researcher On Effects of Overpopulation." *New York Times,*
      Sept. 29, 1995: Obituaries, Accessed Dec. 20, 2010, www.nytimes.com/1995/09/29/obituaries/j-
      b-calhoun-78-researcher-on-effects-of-overpopulation.html
II.   John B. Calhoun, "Death Squared: The Explosive Growth and Demise of a Mouse Population" Proc.
      roy. Soc. Med. Volume 66 January 1973, pp80-88
III. Ramsden, Edmund and Adams, Jon (2008) Escaping the laboratory: the rodent experiment of John B
      Calhoun and their cultural influence. Working papers on the nature of evidence: how well do 'facts'
      travel?, 23/08. Department of Economic History, London School of Economics and Political Science,
      London, UK.

Mouse Universe Population Growth #25

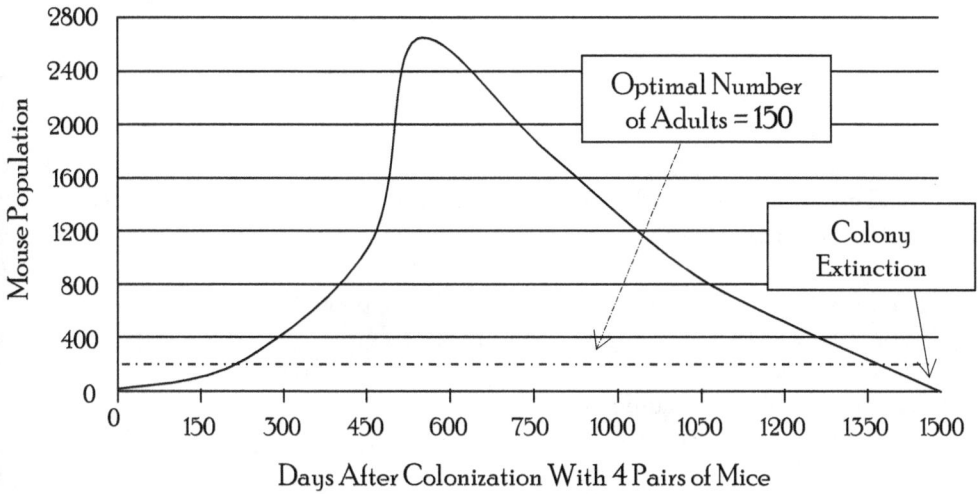

Mouse Population

Optimal Number
of Adults = 150

Colony
Extinction

2800
2400
2000
1600
1200
800
400
0

0 150 300 450 600 750 1000 1050 1200 1350 1500

Days After Colonization With 4 Pairs of Mice

Human Population Growth

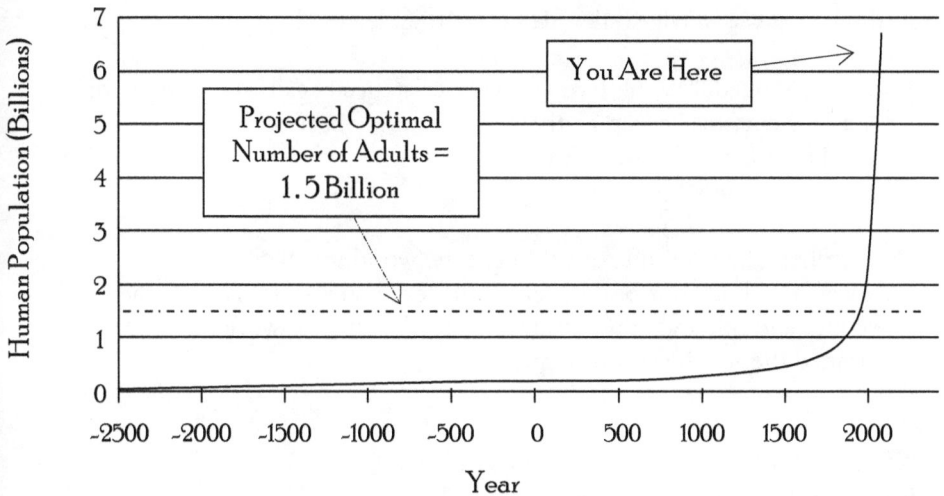

Human Population (Billions)

You Are Here

Projected Optimal
Number of Adults =
1.5 Billion

7
6
5
4
3
2
1
0

-2500 -2000 -1500 -1000 -500 0 500 1000 1500 2000

Year

—Figure 9—

A Calhoun Mouse Universe Population Growth and Decline Graph,
Compared with a Human Population Growth Chart

Philosopher Frog Expands My Mind

I pulled into my driveway simply glad to be alive
and on my front porch I was surprised to find a frog
sitting right where I was about to step.
And given my newfound respect for natural selection and cause and effect,
on a whim, I leaned down, and kissed him.

And the frog looked up at me and said, "Whoa, buddy! Ribit.
You totally just violated my personal space! Ribit.
Now I would have let you kiss me, if you had asked politely.
It is a particularly romantic night. Ribit.
Do you like looking at stars?"

And I replied, "Sure, but I –"
And he interrupted, "Yeah! I like stars! Ribit.
Say, you know about the Big Bang? The beginning of the universe?
Yeah. It really makes me feel small. Ribit!"

And the frog was small, so that was easy to understand.

And he continued, "Yeah! Everything in existence began with a Big Bang!
It's a nice theory backed up by the best scientific observations!
Too bad it's a load of cosmic flycrap!
Ribit.

"Remember when you humans thought the world was flat?
That was based on your best observations too, you just didn't have the
telescopes and other astronomical tools to tell that your worldview was far
too limited the see the complete picture.
Ribit.

"It's like how I used to live in a very small pond.
I thought that pond was all that existed.
Then one day, forces beyond my understanding came along and drained my
pond, so I had to hop off and find a new one.

"In my travels,
I discovered that we actually live on a whole planet covered by ponds!
Thus, I learned that my former view was too limited to see the complete
picture of my universe!

"The same was true with the concept of the flat Earth,
and it will prove true for the Big Bang as well.
Ribit.

"The best deep space observations suggest that all of space and time
exploded from a single point at the beginning of the universe.

"But the truth is, our local universe may have been created by a spacetime
explosion, but that doesn't mean that everything in existence was! Human
history is speckled with conceited scientists foolishly believing that every-
thing they see is everything there is! Like believing the Sun orbits the Earth!

"For all we really know, our universe is just one small pond in an infinite
river of swirling stars and galaxies moving in ebbs and eddies of each other's
influence in an endless symphony of fire and life and destruction and ice!
Yeah, that's what I like thinking about when I look at the stars!
Ribit.
It's beautiful don't you think?"

And I tried to reply,
but my porch morphed into a spacetime ship
and I used the doorknob to steer us to the superluminal expansion black
hole at the beginning of the next universe.

~

Thanks again to Doug Shields for fact-checking & editing this poem!

Senryu

In space,
no one can hear you scream.
Not even if you scream,
for ice cream.

~

Bread.
Mayonnaise.
Cheddar Cheese. Ham. Turkey. Roast Beef. Bacon.
Mustard.
Bread.

~

I folded paper
into our sex.
It is called "origasmi".

~

Hey, get that mind out of the gutter!
I have to stick it back in this guy's head.

~

You have reached the end of side two,
please flip the tape to hear more haiku.

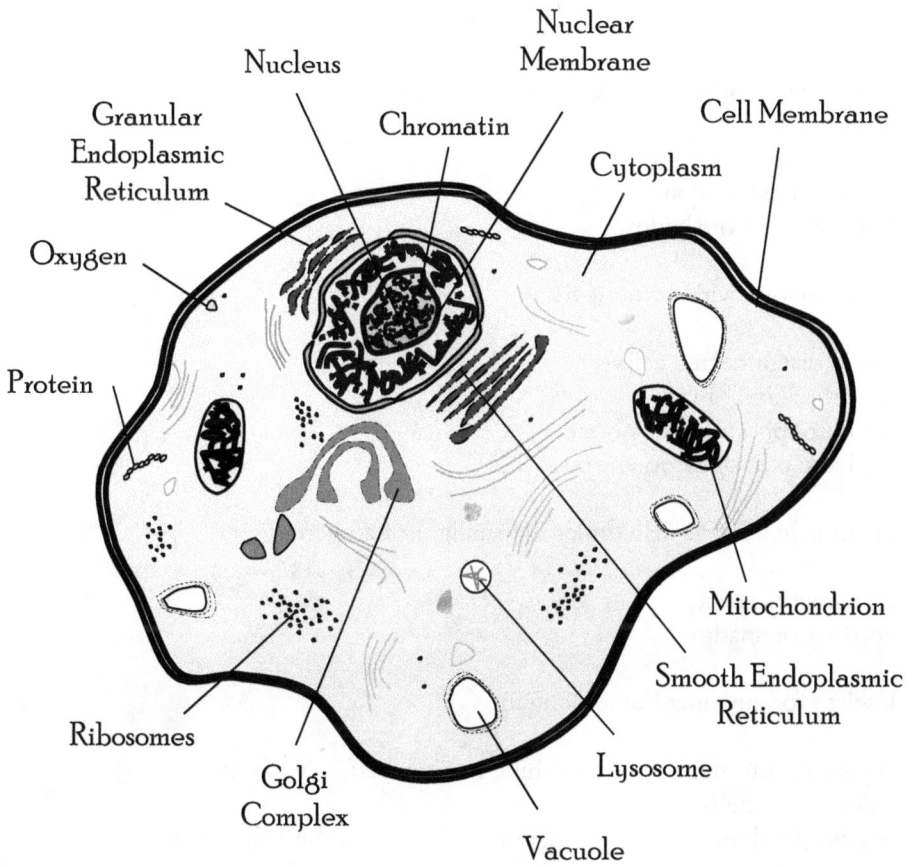

—Figure 10—
A Human Cell
(You Are Here)

An Isolated Cell

What if you could be alone?
With nothing.

In a blank white space.
With no floor or shadow.

A solo phone with no network.

Like a disconnected cell –
the other kind of cell –
the homophone of telephone –
the basic unit of all organisms.

Floating in a Petrie dish under the sunlight of a microscope:

Blank white space,
no floor or shadow.

Useless for anything but looking in.

If you want to think about anything but yourself,
you need some friends,
you need billions of cells to **organize** and form a thinking organ.

And with your new brain,
you might find it fascinating
to pull back the microscope magnification from the petrie dish,
back beyond the scientists' bald spots,
back till the city streets pull away like mouse mazes,
back till the Earth curves out from under you,
smooth and round like a blue skull.

Peel back the atmospheric aura,
and you might find yourself standing in a group of other brain cells,
like neurons poking up out of the convoluted Earth,

and you can feel electricity in the synapse
every time your neighbor gets an epiphany.
Ideas leap between you like flying squirrels between trees.

Or maybe you're more comfortable as the kidneys,
all you ever wanted was to push out waste and restore our cities!

But I'll bet the first cells to organize were called crazy,
said to have faulty circuitry,
just cause they functioned differently.

Like every cell in your body, unique.
Yet somehow each of these entities learned to organize
and form more perfect organs,
which work together to form you.

We could follow our body's example.

If we really want to think,
if we want to solve big problems,
we're gonna need every individual we can get,
even the faulty circuits,
the ones trying to be the blood or the lungs,
the cells dying to be Earth's heart.

But what if every cell in your heart split apart?

What if each just wanted to be alone?

About Danny Strack

Danny Strack is a nationally-known performance poet, and the author of 8 chapbooks of original poetry and artwork.

He was the 2007 Southwest Shootout Regional Individual Champ, a 2008 National Poetry Slam Finalist, 2010 Austin Slam Champion, and is the current Slam Master of the Austin Poetry Slam.

Danny teaches poetry workshops in primary schools and colleges, and his work is frequently used by students in speech competitions.

He is also a technical writer, illustrator, online marketer, maze designer, juggler, burlesque performer and aspiring science fiction author. His primary interests include space-time, humanity, biology, systems theory, technology, comedy, games and poetry.

All graphs and diagrams presented in this book are original creations that are as accurate as possible, based on multiple sources, including, but not limited to those noted on the poems and charts.

For more everything visit: www.dannystrack.com

www.ingramcontent.com/pod-product-compliance
Lightning Source LLC
Chambersburg PA
CBHW061755040426
42447CB00011B/2314